I0486552

THE
MAN
FROM THE
FIFTH DIMENSION

Bruce P. Burns, Ph.D.

Bloomington, IN  Milton Keynes, UK

authorHOUSE®

AuthorHouse™
1663 Liberty Drive, Suite 200
Bloomington, IN 47403
www.authorhouse.com
Phone: 1-800-839-8640

AuthorHouse™ UK Ltd.
500 Avebury Boulevard
Central Milton Keynes, MK9 2BE
www.authorhouse.co.uk
Phone: 08001974150

© 2006 Bruce P. Burns, Ph.D.. All rights reserved.

No part of this book may be reproduced, stored in
a retrieval system, or transmitted by any means
without the written permission of the author.

Revised Edition first published by AuthorHouse 11/1/2006

ISBN: 1-4259-5195-3 (sc)

Library of Congress Control Number: 2006906570

Printed in the United States of America
Bloomington, Indiana

This book is printed on acid-free paper.

Books by the same author

Survival of American Democracy

Abide in Him
(And Be Free)

Also recommended in the same
genre

THE PATH OF LIFE

For a time is coming when people will no longer listen to right teaching. They will follow their own desires and will look for teachers who will tell them whatever they want to hear. They will reject the truth and follow strange myths. (2Timothy 4:3)

I dedicate this to that 'Angel' who is my 'soul mate', who has been with me, and aided me through this entire work.

Unless otherwise indicated, all scripture quotations are taken from the Holy Bible, New Living Translation, copyright 1996. Used by permission of Tyndale House Publisher, Inc., Wheaton, Illinois 60189. All rights reserved.

CONTENTS

O world invisible, we view thee,
O world intangible, we touch thee,
O world unknowable, we know thee,
Inapprehensible, we clutch thee!*

I
THE FIFTH DIMENSION

In these times, many scientists are more
and more giving credence to; 1) the existence
of an intelligent designer, 2) that there may
be reason to believe that there is an unseen
dimension, and 3) that it has far more qualities
and characteristics to its nature than hitherto
suspected.

In life, from time to time one will come across
near geniuses who do not have a lot of common
sense, **or** they have a hidden agenda, such as a
stake. Such is the case of those scientists who
ignore and or deny great advances in human
thought.

One example, among many, was the religious
community's uproar and persecution over the

Copernician <u>theory</u> that the world revolved about the sun, after Galileo promoted it as a <u>fact</u>! Eventually, however, this theory replaced the old one that the earth was the center of the universe!

The rest of this book makes the case that we are at a similar juncture in human events, at this time!

"Today the minds of scientists are reeling as they contemplate the significance of the earth. What is becoming increasingly evident is that the earth is not the center of the universe and not even the center of our solar system. By all logical and mathematical appearances, it seems to be the center of attention for Someone with awesome intelligence and skill who indeed purposed the earth not just for life, but for human life. The facts now asserted about the physics of all the forces and objects in space are compelling scientists to grapple with a theory they refer to as the anthropic principle.

In simple terms, this theory states that by all appearances, the nature and main purpose of the entire universe is to support life, especially human life, on this tiny speck of a planet. They now realize that it is for good reason that the earth is not the center of the universe or of our solar system. If it were, it could not support life. It needs to be exactly where it is, and the entire cosmos must have its exact characteristics for life to exist. Simply put, it takes a universe to make earth the sole 'living' planet." **

Furthermore, electron microscopes reveal the rock solid appearance of this material world as an illusion, and the apparent elusiveness and unsubstantial nature of the spiritual as misleading. For the time being, in order to minimize bias, as much as possible, this will be referred to as the fifth dimension.

Throughout the ages, a few men have demonstrated evidence for the existence of a fifth dimension in various ways and at different times. However, scholars and scientists have failed to access it or even to recognize it. The group among whom a few were able to demonstrate some of its characteristics did not identify it as **the** fifth dimension. Examples that demonstrated the existence of a fifth dimension and illustrated some of its characteristics, all performed by actual historical figures.

1) Ability to control animals,
2) Ability to become impervious to fire,
3) Ability to control natural forces,
4) Ability to communicate in any tongue or language,
5) Ability to heal illnesses, mental, physical and emotional,
6) Ability to predict the future.
7) Ability to see into the fifth dimension and make it visible to others,
8) Ability to defy the physical laws,
9) Ability to be transported instantaneously,
10) Ability to materialize and dematerialize,
11) Ability to reverse or suspend time.
12) Victory over death.

Illustrations and instances of a fifth dimension follow.

* (from, 'In No Strange Land', Francis Thompson)

** RBC Ministries, Discovery Series, Grand Rapids, Mi. 2006 'Celebrating the Wonder of Soil', by Dean Ohlman

II
THE DEN OF LIONS

1-a) so at last the king gave orders for Daniel to be arrested and thrown into the den of lions. The king said to him, "May your God, whom you worship continually, rescue you." A stone was brought and placed over the mouth of the den. The king sealed the stone with his own royal seal and the seals of his nobles, so that no one could rescue Daniel from the lions. Then the king returned to his palace and spent the night fasting. He refused his usual entertainment and couldn't sleep all that night.

Very early the next morning, the king hurried out to the lions' den. When he got there, he called out in anguish, "Daniel, servant of the living God! Was your God, whom you worship continually, able to rescue you from the lions?"

Daniel answered, "Long live the king! My God sent his angel to shut the lions' mouths so that they would not hurt me, for I have been found innocent in his sight. And I have not wronged you, Your Majesty."

The king was overjoyed and ordered that Daniel be lifted from the den. Not a scratch was found on him because he had trusted in his God. Then the king gave orders to arrest the men who had maliciously accused Daniel. He had them thrown into the lions' den, along with their wives and children. The lions leaped on them and tore them apart before they even hit the floor of the den.

Then king Darius sent this message to the people of every race and nation and language throughout the world:

Peace and prosperity to you!

I decree that everyone throughout my kingdom should tremble with fear before the God of Daniel.

For he is the living God,

   And he will endure forever.

His kingdom will never be

destroyed,
   and his rule will never end.
He rescues and saves his people;
   he performs miraculous signs
    and wonders
   in the heavens and on earth.
He has rescued Daniel
   from the power of the lions."

So Daniel prospered during the reign of Darius and the reign of Cyrus the Persian.

III
THE FIERY FURNACE

2-a) Nebuchadnezzar was so furious with Shadrach, Meshach, and Abednego that his face became distorted with rage. He commanded that the furnace be heated seven times hotter than usual. Then he ordered some of the strongest men of his army to bind Shadrach, Meshach, and Abednego and throw them into the blazing furnace. So they tied them up and threw them into the furnace, fully clothed. And because the king, in his anger, had demanded such a hot fire in the furnace, the flames leaped out and killed the soldiers as they threw the three men in! So Shadrach, Meshach, and Abednego, securely tied, fell down into the roaring flames.

But suddenly as he was watching, Nebuchadnezzar jumped up in amazement and exclaimed to his advisers, "Didn't we tie up three men and throw them into the furnace?"

"Yes," they said, "we did indeed, your majesty."

Look! Nebuchadnezzar shouted. "I see four men, unbound, walking around in the fire. They aren't even hurt by the flames! And the fourth looks like a divine being!"

Then Nebuchadnezzar came as close as he could to the door of the flaming furnace and shouted: "Shadrach, Meshach, and Abednego, servants of the Most High God, come out! Come here!" So Shadrach, Meshach, and Abednego stepped out of the fire. Then the princes, prefects, governors, and advisers crowded around them and saw that the fire had not touched them. Not a hair on their head was singed, and their clothing was not scorched. They did not even smell of smoke.

Then Nebuchadnezzar said, "Praise to the God of Shadrach, Meshach, and Abednego! He sent his angel to rescue his servants who trusted in him. They defied the king's command and were willing die rather than serve or worship any

God except their own God. Therefore, I make this decree   : If any people, whatever their race or nation or language, speak a word against the God of Shadrach, Meshach, and Abednego, they will be torn limb from limb, and their houses will be crushed into heaps of rubble. There is no other god who can rescue like this!" Then the king promoted Shadrach, Meshach, and Abednego to even higher positions in the province of Babylon.

IV
WALKING PATH
THROUGH THE SEA

3-a) Then Moses raised his hand over the sea, and the Lord opened up a path through the water with a strong east wind. The wind blew all that night, turning the seabed into dry land. So the people of Israel walked through the sea on dry ground, with walls of water on each side! Then the Egyptians- all of Pharaoh's horses, chariots, and charioteers- followed them across the bottom of the sea. But early in the morning, the Lord looked down on the Egyptian army from the pillar of fire and cloud, and he threw them into confusion. Their chariot wheels began to come off, making their chariots impossible to drive "Let's get out of here!" the Egyptians shouted." "The Lord is fighting for Israel against us!"

When all the Israelites were on the other side, the Lord said to Moses, "Raise you hand over the sea again. Then the waters will rush back over the Egyptian chariots and charioteers." So as the sun began to rise, Moses raised his hand over the sea. The water roared back into its usual place, and the Lord swept the terrified Egyptians into the surging currents. The waters covered all the chariots and charioteers- the entire army of the Pharaoh. Of all the Egyptians who had chased the Israelites into the sea, not a single one survived.

The people of Israel had walked through the middle of the sea on dry land, as the water stood up like a wall on both sides. This was how the Lord rescued Israel from the Egyptians that day. And the Israelites could see the bodies of the Egyptians washed up on the shore. When the people of Israel saw the mighty power that the Lord had displayed against the Egyptians, they feared the Lord and put their faith in him and his servant Moses.

V
SPEAKING IN TONGUES

4-a) On the day of Pentecost, seven weeks after Jesus' resurrection, the believers were meeting together in one place. Suddenly, there was a sound from heaven like the roaring of a mighty windstorm, in the skies above them, and it filled the house where they were meeting. Then, what looked like tongues of fire appeared and settled on each one of them. And everyone present was filled with the Holy Spirit and began speaking in other languages, as the Holy Spirit gave them this ability. Godly Jews from many nations were living in Jerusalem at that time. When they heard this sound they came running to see what it was all about, and they were bewildered to hear their own languages being spoken by the believers.

They were beside themselves with wonder. "How can this be?" they exclaimed. "These people are all from Galilee, and yet we hear them speaking the languages of the lands where we were born! Here we are – Parthians, Medes, Elamites, people from Mesopotamia, Judea, Cappadocia, Pontus, the province of Asia, Phrygia, Pamphaylia , Egypt, and the areas of Libya toward Cyrene, visitors from Rome (both Jews and converts to Judaism), Cretans, and Arabians. And we hear all these people speaking in our own languages about the wonderful things God has done!" They stood there amazed and perplexed. "What can this mean?" they asked each other. But others in the crowd were mocking. "They're drunk, that's all!" they said.

Then Peter stepped forward with the eleven other apostles and shouted to the crowd, "Listen carefully, all of you fellow Jews and residents of Jerusalem! Make no mistake about this. Some of you are saying these people are drunk. It isn't true! It's much too early for that. People don't get drunk by nine o'clock in the morning. No, what you see this morning was predicted centuries ago by the prophet Joel:

'In the last days, God said,
 I will pour out my Spirit upon all
  People
Your sons and daughters will prophesy,
 your young men will see visions,
 and your old men will dream
  dreams.
In those days I will pour out my Spirit
 upon all my servants, men and
  women alike,
 and they will prophesy.
And I will cause wonders in the
  heavens above
 and signs on the earth below –
 blood and fire and clouds of smoke.
The sun will be turned into darkness'
 and the moon will turn bloodred,
 before the great and glorious day of
  the Lord arrives.
And anyone who calls on the name of
  the Lord
 will be saved.

VI
THE BLIND SEE

5-a) As Jesus was walking along, he saw a man who had been blind from birth. "Teacher," his disciples asked him, "why was this man born blind? Was it a result of his own sins or those of his parents?"

"It was not because of his sins or his parents' sins," Jesus answered. "He was born blind so the power of God could be seen in him. All of us must quickly carry out the tasks assigned to us by the one who sent me, because there is little time left before the night falls and all work comes to an end. But while I am still here in the world, I am the light of the world."

Then he spit on the grounds making mud with the saliva, and smoothed the mud over the blind man's eyes. He told him, "Go wash in the pool of Siloam." (Siloam means Sent). So the man went and washed, and came back seeing!

His neighbors and others who knew him as a blind beggar asked each other, " Is this the same man – that beggar?" Some said he was, and others said, "No, but he surely looks like him!"

And the beggar kept saying, "I am the same man!"

They asked, "Who healed you? What happened?"

He told them, "The man they call Jesus made mud and smoothed it over my eyes and told me, 'Go to the pool of Siloam and wash off the mud.' I went and washed, and now I can see!"

"Where is he now, they asked?"

"I don't know," he replied.

Then they took the man to the Pharisees. Now as it happened, Jesus had healed the man on a Sabbath. The Pharisee asked the man all about it. So he told them, "He smoothed the mud over my eyes, and when it was washed away, I could see!"

Some of the Pharisees said, "This man Jesus is not from God, for he is working on the Sabbath." Others said, "But how could an ordinary sinner do such miraculous signs?" So there was a deep division of opinion among them.

Then the Pharisees once again questioned the man who had been blind and demanded, "This man who opened your eyes- who do you say he is?"

The man replied, "I think he must be a prophet."

The Jewish leaders wouldn't believe he had been blind, so they called in his parents. They asked them, "Is this your son? Was he born blind? If so how can he see?"

His parents replied, "We know this is our son and that he was born blind, but we don't know how he can see or who healed him. He is old enough to speak for himself. Ask him." They said this because they were afraid of the Jewish leaders, who had announced that anyone saying

Jesus was the Messiah would be expelled from the synagogue. That's why they said, "He is old enough to speak for himself. Ask him."

So for the second time they called in the man who had been blind and told him, "Give glory to God by telling the truth, because we know Jesus is a sinner."

"I don't know whether he is a sinner," the man replied. "But I know this: I was blind, and I can see!"

"But what did he do?" they asked. "How did he heal you?"

"Look!" the man exclaimed. "I told you once. Didn't you listen? Why do you want to hear it again? Do you want to become his disciples, too?"

Then they cursed him and said, "You are his disciple, but we are disciples of Moses. We know God spoke to Moses, but for this man, we don't know anything about him."

"Why, that's very strange!" the man replied. "He healed my eyes, An yet you don't know anything about him!" Well, God doesn't listen to sinners, but he is ready to hear those who worship him and do his will. Never since the world began has anyone been able to open the eyes of someone born blind. If this man were not from God, he couldn't do it."

"You were born in sin!" they answered. "Are you trying to teach us?" And they threw him out of the synagogue..

When Jesus heard what had happened, he found the man and said, "Do you believe in the Son of Man?"

The man answered, "Who is he, sir because I would like to."

"You have seen him," Jesus said, "and he is speaking to you!"

"Yes, Lord," the man said, "I believe!" And he worshipped Jesus.

Then Jesus told him, "I have come to judge the world. I have come to give sight to the blind and to show those who think they see that they are blind."

The Pharisees who were standing there heard him and asked' "Are you saying we are blind?"

"If you were blind, you wouldn't be guilty," Jesus replied. " But you remain guilty because you claim you can see."

VII
INTERPRETING DREAMS

6-a)   Pharaoh sent for Joseph at once, and he was brought hastily from the dungeon. After a quick shave and change of clothes, he went in and stood in the Pharaoh's presence. "I had a dream last night," Pharaoh told him, "and none of these men can tell me what it means. But I have heard that you can interpret dreams, and that is why I have called for you."

"It is beyond my power to do this," Joseph replied. "But God will tell you what it means and will set you at ease."

So Pharaoh told him the dream. "I was standing on the bank of the Nile River," he said. "Suddenly, seven fat, healthy-looking cows came up out of the river and began grazing along its bank.   But then seven other cows came up from the river. They were very thin and gaunt-in fact, I've never seen such ugly animals in all the land of Egypt. These thin, ugly cows ate up the seven fat ones that had come out of the river first, but afterward they were still as ugly and gaunt as before! Then I woke up.

A little later I had another dream. This time there were seven heads of grain on one stalk, and all seven heads were plump and full. Then out of the same stalk came seven withered heads, shriveled by the east wind. And the withered heads swallowed up the plump ones! I told these dreams to my magicians, but not one of them could tell me what they mean."

"Both dreams mean the same thing," Joseph told the Pharaoh. "God was telling you what he is about to do. The seven fat cows and the seven plump heads of grain both represent seven years of prosperity. The seven thin, ugly cows and the seven withered heads of grain represent seven years of famine. This will happen just as I have described, for God has shown you what he is about to do. The next seven years will be a period of great prosperity throughout the land of Egypt. But afterward there will be seven years of famine so great that all the prosperity will be forgotten and wiped out. Famine will destroy the land. This famine will be so terrible that even the memory of the good years will be erased. As for having the dream twice, it means that the matter has been decreed by God and that he will make these events happen soon.

My suggestion is that you find the wisest man in Egypt and put him in charge of a nationwide program. Let Pharaoh appoint officials over the land, and let them collect one-fifth of all the crops during the seven good years. Have them gather all the food and grain of these good years into the royal storehouses, and store it away so there will be food in the cities. That way there will be enough to eat when the seven years of famine come. Otherwise disaster will surly strike the land, and all the people will die."

Joseph's suggestions were well received by Pharaoh and his advisers. As they discussed who should be appointed for the job, Pharaoh said, "Who could do it better than Joseph? For he is a man who is obviously filled with the spirit of God." Turning to Joseph, Pharaoh said, "Since God has revealed the meaning of the dreams to you, you are the wisest man in the land! I hereby appoint you to direct this project. You will manage my household and organize my people. Only I will have a rank higher than yours."

And Pharaoh said to Joseph, "I hereby put you in charge of the entire land of Egypt." Then Pharaoh placed his own signet ring on Joseph's finger as a symbol of his authority. He dressed him in beautiful clothing and placed the royal gold chain about his neck. Pharaoh also gave Joseph the chariot of his second-in-command, and where ever he went the command was shouted, "Kneel Down!" So Joseph was put in charge of all Egypt. And Pharaoh said to Joseph, "I am the king, but no one will move a hand or a foot in the entire land of Egypt without your approval."

Pharaoh renamed him Zaphe nath-paneah and gave him a wife – a young woman named Asenath, the daughter of Potiphera, priest of Heliopolis. So Joseph took charge of the entire land of Egypt. He was thirty years old when he entered the service of Pharaoh, the king of Egypt. And when Joseph left Pharaoh's presence, he made a tour of inspection throughout the land.

And sure enough, for the next seven years there were bumper crops everywhere. During those years, Joseph took a portion of all the crops grown in Egypt and stored them for the government in nearby cities. After seven years, the granaries were filled to overflowing. There was so much grain, like sand on the seashore, that the people could not keep track of the amount.

During this time, before the arrival of the first of the famine years, two sons were born to Joseph and his wife, Asemath, the daughter of Potiphera, priest of Heliopolis. Joseph names his older son Manasseh, for he said, God has made me forget all my troubles and the family of my father." Joseph named his second son Ephraim, for he said, "God has made me fruitful in this land of my suffering."

At last the seven years of plenty came to an end. Then the seven years of famine began, just as Joseph had predicted. There were crop failures in all the surrounding countries, too, but in Egypt there was plenty of grain in the storehouses. Throughout the land of Egypt the people began to starve. They pleaded with Pharaoh for food, and he told them, "Go to Joseph and do whatever he tells you." So with severe famine everywhere in the land, Joseph opened up the storehouses and sold grain to the Egyptians. And people from surrounding lands also came to Egypt to buy grain from Joseph because the famine was severe throughout the world.

VIII
SEEING THE INVISIBLE

7-a) When the king of Aram was at war with Israel, he would confer with his officers and say, "We will mobilize our forces at such and such a place."

But immediately Elisha, the man of God, would warn the king of Israel, "Do not go near that place, for the Arameans are planning to mobilize their troops there." So the king of Israel would send word to the place indicated by the man of God, warning the people there to be on their guard. This happened several times.

The king of Aram became very upset over this. He called in his officers and demanded, "Which of you is the traitor? Who has been informing the king of Israel of my plans?"

It's not us, my lord," one of the officials replied. "Elisha, the prophet in Israel, tells the king of Israel even the words you speak in the privacy of your bedroom!"

The king commanded, "Go and find out where Elisha is, and we will send troops to seize him."

And the report came back: "Elisha is at Doshan." So one night the king of Aram sent a great army with many chariots and horses to surround the city. When the servant of the man of God got up early the next morning and went outside, there were troops, horses, and chariots everywhere.

"Ah, my lord, what will we do now?" he cried out to Elisha.

"Don't be afraid!" Elisha told him. "For there are more on our side than on theirs!" Then Elisha prayed, "O Lord, open his eyes and let him see!" The Lord opened his servant's eyes, when he looked up, he saw that the hillside around Elisha was filled with horses and chariots of fire.

IX

A BANQUET
FOR THOUSANDS
ON FIVE LOAVES

8-a) That evening the disciples came to him and said, "This is a desolate place, and it is getting late. Send the crowds away so they can go to the villages and buy food for themselves."

But Jesus replied, "That isn't necessary – you feed them."

"Impossible!" they exclaimed. "We have only five loaves of bread and two fish!"

"Bring them here," he said. Then he told the people to sit down on the grass. And he took the five loaves and two fish, looked up toward heaven, and asked God's blessing on the food. Breaking the loaves into pieces, he gave some of the bread and fish to each disciple, and the disciples gave them to the people. They all ate as much as they wanted, and they picked up twelve baskets of leftovers. About five thousand men had eaten from those five loaves, in addition to all the women and children!

Immediately after this, Jesus made his disciples get back into the boat and cross to the other side of the lake while he sent the people home. Afterward he went up into the hills by himself to pray. Night fell while he was there alone. Meanwhile, the disciples were in trouble far away from land, for a strong wind had risen, and they were fighting heavy waves.

About three o'clock in the morning Jesus came to them, walking on the water. When the disciples saw him, they screamed in terror, thinking he was a ghost.

But Jesus spoke to them at once. "It's all right," he said. "I am here! Don't be afraid."

Then Peter called to him, "Lord, if it's really you, tell me to come to you by walking on the water.

"All right, come," Jesus said."

So Peter went over the side of the boat and walked on the water to Jesus. But when he looked around at the high waves, he was terrified and began to sink. "Save me, Lord! he shouted.

Instantly Jesus reached out his hand and grabbed him. "You don't have much faith," Jesus said. "Why did you doubt me?" And when they climbed back into the boat, the wind stopped.

Then the disciples worshipped him. "You really are the Son of God!" they exclaimed. The news of their arrival spread quickly throughout the whole surrounding area, and soon people were bringing all their sick to be healed. The sick begged him to let them touch even the fringe of his robe, and all who touched it were healed.

X
Instant Transportation

9-a) As for Philip, an angel of the Lord said to him, "Go south down the desert road that runs from Jerusalem to Gaza." So he did, and he met the treasurer of Ethiopia, a eunuch of great authority under the queen of Ethiopia. The eunuch had gone to Jerusalem to worship, and he was now returning. Seated in his carriage, he was reading aloud from the book of the prophet Isaiah.

The Holy Spirit said to Philip, "Go over and walk along beside the carriage."

Philip ran over and heard the man reading from the prophet Isaiah; so he asked, "Do you understand what you are reading?"

The man replied, "How can I, when there is no one to instruct me?" And he begged Philip to come up into the carriage and sit with him.

The passage of Scripture he had been reading was this:

"He was led as a sheep to the slaughter:

And as a lamb is silent before the
Shearers,

he did not open his mouth.

He was humiliated and receive no
justice.

Who can speak of his descendants?

For his life was taken from this earth."

The eunuch asked Philip, "Was Isaiah talking about himself or someone else?" So Philip began with this same Scripture and then used many others to tell him the Good News about Jesus.

As they rode along, they came to some water, and the eunuch said, "Look! There's some water! Why can't I be baptized?" He ordered the carriage to stop, and they went down into the water; and Philip baptized him.

When they came up out of the water, the Spirit of the Lord caught Philip away. The eunuch never saw him again but went on his way rejoicing. Meanwhile, Philip found himself farther north at the city of Azotus! He preached the Good News there and in every city along the way until he came to Caesarea.

XI
DEMATERIALIZATION AND MATERIALIZATION

10-a) That evening, on the first day of the week, the disciples were meeting behind locked doors because they were afraid of the Jewish leaders. Suddenly, Jesus was standing there among them! "Peace be with you," he said. As he spoke, he held out his hands for them to see, and he showed them his side. They were thrilled with joy when they saw their Lord! He spoke to them again and said, "Peace be with you. As the Father has sent me, so I send you." Then he breathed on them and said to them, "Receive the Holy Spirit. If you forgive anyone's sins, they are forgiven. If you refuse to forgive them, they are unforgiven."

One of the disciples, Thomas (nicknamed the Twin), was not with the others when Jesus came. They told him, "We have seen the Lord!

But he replied, "I won't believe it unless I see the nail wounds in his hands, put my fingers into them, and place my hand into the wound in his side."

Eight days later the disciples were together again, and this time Thomas was with them. The doors were locked; but suddenly, as before, Jesus was standing among them. He said, "Peace be with you." Then he said to Thomas, "Put your finger here and see my hands. Put your hand into the wound in my side. Don't be faithless any longer. Believe!"

"My Lord and my God!" Thomas exclaimed.

Then Jesus told him, "You believe because you have seen me. Blessed are those who haven't seen me and believe anyway."

XII
THE SUN STANDS STILL

11-a) The men of Gibeon quickly sent messengers to Joshua at Gilgal, "Don't' abandon your servants now!" they pleaded. "Come quickly and save us! For all the Amorite kings who live in the hill country have come out against us with their armies."

So Joshua and the entire Israelite army left Gilgal and set out to rescue Gibeon. "Do not be afraid of them," the Lord said to Joshua, "for I will give you victory over them. Not a single one of them will be able to stand up to you."

Joshua traveled all night from Gilgal and took the Amorite armies by surprise. The Lord threw them into a panic, and the Israelites slaughtered them in great numbers at Gibeon. Then the Israelites chased the enemy along the road to Bethhoron and attacked them at Azekah and Makkedah, killing them along the way. As the

Amorites retreated down the road from Beth-horon, the Lord destroyed them with a terrible hailstorm that continued until they reached Azekah. The hail killed more of the enemy than the Israelites killed with the sword.

On the day the Lord gave the Israelites victory over the Amorites, Joshua prayed to the Lord in front of all the people of Israel, He said,

Let the sun stand still over Gibeon,

And the moon over the valley of Aijalon.

So the sun and moon stood still until the Israelites had defeated their enemies.

Is this event not recorded in *The Book of Jashar*?" The sun stopped in the middle of the sky, and it did not set as on a normal day. The Lord fought for Israel that day. Never before or since had there been a day like that one, when the Lord answered such a request from a human being. Then Joshua and the Israelite army returned to their camp at Gilgal.

XIII
Rising From the Dead

12-a) The next day- on the first day of the Passover ceremonies – the leading priests and Pharisees went to see Pilate. They told him, "Sir, we remember what that deceiver once said while he was still alive: 'After three days I will be raised from the dead.' So we request that you seal the tomb until the third day. This will prevent his disciples from coming and stealing his body and then telling everyone he came back to life! If that happens, we'll be worse off than we were at first."

Pilate replied, "Take guards and secure it the best you can." So they sealed the tomb and posted guards to protect it.

Early on Sunday morning, as the new day was dawning, Mary Magdalene and the other Mary went out to see the tomb. Suddenly there was a great earthquake, because an angel of the Lord came down from heaven and rolled aside the stone and sat on it. His face shone like lightning, and his clothing was as white as snow. The guards shook with fear when they saw him, and fell into a dead faint.

Then the angel spoke to the women. "Don't be afraid!" he said. "I know you are looking for Jesus, who was crucified. He isn't here! He has been raised from the dead, just as he said would happen. Come; see where his body was lying. And now, go quickly and tell his disciples he has been raised from the dead, and he is going ahead of you to Galilee. You will see him there. Remember, I have told you."

The women ran quickly from the tomb. They were very frightened but also filled with great joy, and they rushed to find the disciples to give them the angel's message. And as they went, Jesus met them. "Greetings!" he said. And they ran to him, held his feet, and worshiped him. Then Jesus said to them, "Don't be afraid1 Go tell my brothers to leave for Galilee, and they will see me there."

As the women were on their way into the city, some of the men who had been guarding the tomb went to the leading priests and told them what had happened. A meeting of all the religious leaders was called, and they decided to bribe the soldiers. They told the soldiers, "You must say, 'Jesus' disciples came during the night while we were sleeping, and they stole his body.' If the governor hears about it, we'll stand up for you and everything will be all right." So the guards accepted the bribe and said what they were told to say. Their story spread widely among the Jews, and they still tell it today.

References for the quotations on pages 6 through 43 are:

[1-a Daniel 6:16-23]   [7-a 2 Kings 6:8-17]

[2-a Daniel 3:19-30]   [8-a Matthew 14:22-36]

[3-a Exodus 14:21-31]   [9-a Acts 8:26-40]

[4-a Acts 2:1-21]   [10-a John 20:19-29]

[5-a John 9:1-41]   [11-a Joshua 10:6-15]

[6-a Genesis 41:14-57]

[12-a Matthew 27:62-66 & 28:1-15]

Does the fish soar to find the ocean,
The eagle plunge to find the air-
That we ask of the stars in motion
If they have rumor of thee there?*

XIV
CALLED FOR THE FIFTH DIMENSION

We have listed some characteristics and occurrences of the fifth dimension. We have not, however related how and whether we can avail ourselves of its qualities and whether a person could enter the fifth dimension.

The fifth dimension is not accessed by mathematics, philosophy, economics, physics, chemistry, astronomy, séances or magic. In fact, many scientists deny its existence because it is not in keeping with the physical laws of the known universe. For various reasons some choose not to give credence to it. This proves a truism – 'none are as blind as those who **will** not see'. In other words, if you do not want to believe something you cannot.

The world, its laws and order, as we know it, scream out for an Intelligent Creator. To try to explain away the Creator by arguments such as; a **theory** like Evolution, or by accident; takes such a stretch of imagination that it would only be possible for a person to do so by:

1) relying solely on the physical senses, and then only by choosing not to give any honest acknowledgement as to the reason for the order and the dependable laws of the universe;

2) or by making up an intellectual rationalization such as evolution or mere accident and denying any arguments or 'proofs' to the contrary, most likely because of some vested interest!

There is one thing that is needed for one to benefit from continued reading. It would be helpful for the reader to keep an open mind, so as not to jump too quickly to unnecessary or untrue conclusions. This may be a first step toward being able to utilize the fifth dimension!

Some of the more renowned men who have accessed the fifth dimension were- Enoch, Abraham, Jacob, Joseph, Moses, Joshua, David, Elijah, Elisha, Daniel, Isaiah, Jeremiah, John, Peter, and Paul.

These stories were all true and many of them were prophesied (predicted) centuries before they came about. In fact, this feature is verification of the truth of these stories. There are still many prophesies yet to be fulfilled!

The same source depicts both past and future events. In addition, it also describes human nature revealing both good and bad aspects, of it. It predicts how man is going to deal with this world.

God is not going to inspire us unless we are able to react to His Word (Bible). Faith is a gift from God; if we do react to His Word then we are capable of receiving His breath of faith. Even if that first reaction is unbelief, we have been inspired if that unbelief is followed by curiosity, interest, or being challenged. This is shown by interest and study to learn more or an effort to prove or disprove His Word.

If you continue to read, then you are surely 'called'. Whether you agree or disagree is another matter. A positive confirmation of being either 'chosen' or rejected will be a matter confirmed within you.

Entering the Fifth Dimension does not depend upon money, property, or wealth of any kind; intelligence, knowledge, or even wisdom; fame, or popularity; nor power of any kind. In fact, an overabundance of any one of them may be a detriment.

Entering the Fifth Dimension depends upon common sense, old-fashioned savvy, and, faith. Only if faith is combined with the five senses of: hearing, smelling, tasting, seeing, and touching which God gave to humankind; can we 'see' God in His awesomeness. This is displayed in nature, and in the order and laws of the Universe. However, **faith** is a gift from God and is given only to the humble and the seeking!"

Is it possible for us to enter the Fifth Dimension? Unbelievably, the answer is yes! However, how does one? In all of recorded history, there has been only one human to come from the fifth dimension and return to it.

It would be very helpful to learn more about this man who not only demonstrated many of its characteristics but who also came from and returned to this dimension. It is said that more books have been written about him than any other single person who has ever lived.

Through faith, we must surpass having only physical or material, orientations, because we are seeking the properties and characteristics of a special dimension, the Fifth Dimension!

The generic term that encompasses The Fifth Dimension is Spiritual. It is accessed through faith and agape love. (Agape love means, self-giving loyal concern that freely accepts another and seeks his or her good). The 'source', mentioned at the outset, is the Bible and, the Race of People was the Jews.

If the question arises, 'Have I been chosen?' the odds are in your favor. The confirmation is in two parts. **

1) If any of the below is true for you – you have been **called**:

   a) You want to or do believe Jesus Christ is the Son of God &/or

   b) You ponder, what is the meaning of life, why am I here? &/or

   c) You want or are seeking faith or religion
       &/or

   d) You are convinced that there is intelligent design to the Universe.

However, that you have been **chosen**

2) Is only shown (**not determined**) by your works and your life,

In other words: Your works are the good **you do** **to and for others** (shows how obedient you are to God's Ten Commandments); your life however, are your thoughts, words, goals, emotions, and motives (**<u>reveals how much</u> <u>you</u> <u>are led</u>** by the Holy Spirit.)

All of this is only a confirmation of your salvation. Salvation is not a result of your efforts. Your faith in and reliance upon Jesus Christ is the only reason for your salvation. Your hope for and assurance of salvation are your works and your life. Thus, your life is to be based <u>upon your hope of salvation.</u> That is why **hope is rated right up there with faith and love.** (See 1 Corinthians 13:13)

* (from, 'In No Strange Land', by Francis Thompson)

[** My authority for parts one and two on page 49 is; Matthew 7:7-8, 1Timothy 2:4-6, and 1Timothy 1:15 …"] **(In the above, to ask is to want, to seek is to try, and to knock is to do)** [<u>To be given is to be called, to find is to be blessed with faith, to be opened is to enter the Fifth Dimension</u>)

Not where the wheeling systems darken,
And our benumbed conceiving soars!-
The drift of pinions, would we hearken,
Beats at our own clay-shuttered doors.*

XV
CHOSEN FOR THE FIFTH DIMENSION

The greatest practitioner of The Fifth Dimension, and the only man to come from, to demonstrate, and return to The Fifth Dimension was Jesus Christ - The Son of God. The Bible contains all the information one needs to learn about more of its characteristics. Most importantly, it reveals how anyone can realistically gain permanent residence in the wonderful world of the fifth.

Why is it so hard for some people to turn their unbelief into faith? According to The Bible, it was pride, which was man's first and original sin when he disobeyed God in the Garden of

Eden. Why was this so fundamental? Because it takes faith in God to prevent pride.

In other words, pride is of humans only and even though it was the basic sin, the fundamental correction for everything is **faith,** which is from God.

A cardinal and basic law has always been, "… it is impossible to please God without faith. Anyone who wants to come to Him must believe that there is a God and that He rewards those who sincerely seek Him." (Hebrews 11:6)

The 'pride' and lack of faith, rested in the fact that Adam and Eve believed the devil when he told them, in so many words, that God had lied to them and that if they disobeyed Him, they would become like God. Thereupon they disobeyed God and ate the forbidden fruit.

God chose us, *before we were even born,* that we <u>could</u> (not would), become His children, sons and daughters. We were thus, predestinated for the fifth dimension, But why were we predestinated, - because we were unable to earn it!

Therefore, since we have no room for pride, that is, no one has nor can enter the fifth dimension because of their own efforts or worth alone. The glory is all to be God's glory.

This allows us to realize that God's Ten Commandments alone cannot convert the *sinner* to sainthood! It is solely through His grace alone! Thus demonstrating that God is our Creator and we are His creatures and that **all** the glory belongs to Him!

This means that if you have been predestinated for the fifth dimension, you can eventually qualify for that simply because God gifted you with faith, making it possible.

In other words, that good seed which God planted in you before birth can eventually crowd out the 'weeds' which you have allowed to grow in your life. If your true manifest destiny will triumph – it is to God's glory, not yours. You will eventually understand this is the case and will love to glorify God!

His laws of salvation are summed up: 1) love the Lord your God with all your heart, all your soul, and all your mind; 2) love your neighbor as yourself. Thus, they are positive and not negative commands!

Some think that God chose us because of His foreknowledge. That is, He knew in advance, what our decisions would be and that we would eventually choose to do His Will, on our own! If we did not accomplish this on our own, we would be denied entrance to the fifth dimension. This thinking, however, leads to pride because in essence we have thus accomplished entrance into the fifth dimension by our own efforts.

Because of Adam and Eve's sin (of pride), the entire human race has sin as a part of its' genetic make up. No matter how good a human tries to be, he/she will never be perfect because of that flaw in us. Therefore, one must accept Christ's perfection as his/her substitute.

If proof of this pride is needed, just ask yourself. Why is it so difficult to accept the fact that only Christ is a perfect human and only He can substitute for anyone who wants Heaven rather than Hell? Thanks are to God for loving humans so much that He gave His only begotten Son, as the perfect sacrifice!

That is, we must use the faith He has given us. In other words, we must believe: a) Jesus is the Son of God, and b) that we accept Him as our savior and our necessary redeemer. The only proof we have done this is our life and our works. Upon which we shall be judged to be worthy of certain rewards or 'crowns' as Paul put it, in the Fifth Dimension

To help accomplish the foregoing we need to study the Bible, commune with other Christians, and pray. We all are sinners and we will not be perfect until we are united with Jesus after He returns for us.

To repeat. It is important to keep in mind that we are saved by Jesus Christ, we do not earn salvation nor do we deserve it. HOWEVER, **our spiritual rewards** in and of the Fifth Dimension may *depend both upon our works and our life here on earth*. **

Our works and our life consist of loving God and loving others (all humans, to the best of our ability). Salvation is from believing in Christ and trusting Him. This means basing our lives on Him. (He is to be our way, our truth, and our life), and acknowledging Him as the Son of God, and the *only means of our salvation*.

God is supreme, sovereign, overall, the source, creator, and originator of everything, the alpha and the omega, the first and last. Humans are his creatures and have been created to glorify Him.

Other gifts from God (in addition to faith and salvation) are Jesus Christ, the Holy Spirit, The Bible, life, love, and truth. The final gift is to be in the Fifth Dimension, which means to have eternal life in Heaven; with Almighty

God Who is the Father, the Son, and the Holy Spirit.

Before one judges another person as to whether or not that person is or will be saved, let him keep in mind what Saint Paul said, "God, wants everyone to be saved and to understand the truth..." (1Timothy 2:4)

* (from, 'In No Strange Land' by Francis Thompson)

** Works or deeds always implies being done with a loving attitude and faith.

The angels keep their places;-
Turn but a stone, and start a wing!
'Tis ye, 'tis your estranged faces,
That miss the many-splendored thing.*

XVI
MAN OR GOD?

Who was the man from the Fifth Dimension?
He it is that demonstrated many characteristics
of that dimension such as:

1) - control the forces of nature, such as the
   wind and the waves,
2) - heal illnesses,
3) - predict the future,
4) - defy physical laws,
5) - be transported instantly;
6) - materialize and dematerialize,
7) - raised the dead;
8) - casts out demons from people;
9) - feed thousands with a few loaves and a
   few fish,
10) - was himself resurrected from the dead.

Others had previously:

11) - been impervious to fire;

12) - controlled wild ferocious animals;

13) - communicated in unknown languages and tongues;

14) - seen into the Fifth Dimension and made it visible to others,

15) - reversed and suspended time.

"I don't know Who- or what- put the question, I don't know when it was put. I don't even remember answering. But at some moment I did answer YES to Someone – or Something- and from that hour I was certain that existence is meaningful and that, therefore, my life, in self-surrender, had a goal." (Markings by Dag Hammerskjold)

That 'Someone or Something' Dag Hammerskjold referred to is Jesus Christ. His statement is reminiscent of Saint Paul's statement to the Greeks: "Men of Athens, I notice that you are very religious, for as I was walking along I saw your many altars. And one of them had this inscription on it – 'To an Unknown God.' (Acts 17:22-23).

As stated previously, Jesus Christ was and is that one person to have come from and returned to the Fifth Dimension, which is Heaven. His life demonstrated; love, joy, peace, patience, kindness, gentleness, goodness, humility, faith, self-denial, healing and all kinds of miracles

He has made all this available to us by sending to us His Holy Spirit. The degree to which we display all of these qualities and abilities reveals the quality and quantity of our faith.

Christ lived a life of joy and tears, joy for his Apostles and future believers in Him and tears for the world and those who rejected (or ignored) Him. In many places, Christ said 'I did not come for the righteous but for sinners'. Paul confirmed this in Romans 3:10-12, as did Psalms 14:2-3.

The righteous are self-righteous because as Isaiah said, (Isaiah 64:6) "…our righteousnesses are as filthy rags..." Paul said, 'Christ is our righteousness.' In the Psalms it was said,' there are none good, no not one.'

Therefore, if anyone feels righteous without Christ he must be feeling, 'self-righteous'. (This was much the same as the Pharisees and Sadducees who rejected Christ felt).

Christ came for all men, as all men are sinners. Those who felt righteous felt that they did not need Him and thus rejected Him. God wants *all men* to be saved. Any person who is self-righteous does not feel the need for Christ. Such a person thus dismisses Him, or uses Him as a front (i.e., they do not believe what they preach) and therefore are beyond the reach of God because God has given us free choice.

To be a little more explicit about what saves a person, belief in Christ or a good life with good works. This is a confusing area. Only turning your life over to Christ gives a person salvation (eternal life in Heaven).

However, what does turning your life over to Christ or believing in Him mean? It means we want to please Him, obey Him, purify ourselves, to be like Him; otherwise where is the proof we really believe in Him?

Believing in Christ is the essential requirement but that may only be self-deception, unless our life is the confirmation of this. (Authority, 1 John 3:2*ff*; there are many other times and in different ways this same thought is expressed in the Bible).

In other words, it takes both trust and faith in Jesus Christ and a good life (or at least a valiant attempt at it). These two factors are the only sufficient **and** convincing sign of this. see Rev. 20:12

To quote Romans 6:5-11: "Since we have been united with Him in His death, we will also be raised as He was. Our old sinful selves were crucified with Christ so that sin might lose its power in our lives. We are no longer slaves to sin. For when we died with Christ, we were set free from the power of sin. In addition, since we died with Christ, we know we will also share His new life. We are sure of this because Christ rose from the dead, and He will never die again. Death no longer has any power over Him. He died once to defeat sin, and now He lives for the glory of God. So you should consider yourselves dead to sin and able to live for the glory of God through Christ Jesus."

It is important to remember also that we will never be perfect as long as we live on the earth. Not until Christ returns for us will we have a glorified body and a spiritual mind. Until then we just have to keep trying. Some of the many references to this in the New Testament are Philippians 3:12-15; Ephesians 4:12*ff*; John 17:23; 2 Corinthians 7:1, & 12:9.

Jesus Christ was God made Man, more on this in Chapter 21

* (from, 'In No Strange Land' by Francis Thompson)

Let not ambition mock their useful toil,
Their homely joys and destiny obscure;
Nor grandeur hear with a disdainful smile
The short and simple annals of the poor. *

XVII
THE SELF

Another important thing to consider for a 'would-be' Christian is the **SELF**.

The *'self'* is an insidious interloper. How is this so? Because it begins at birth and is an effort by Satan to replace God at a person's center. This is borne out by the way the world (which is Satan's domain) looks at self-confidence, self-esteem, and self-worth, for example; of which the 'burn-out' side is, selfishness, self-centeredness, self-consciousness.

Christ asked of us only three things: 1) to deny our-self, 2) to take up our cross, and 3) to follow Him! This is one of those triads that are so important; for example, on the positive side is faith, hope, and love. The ultimate triad is,

God, the Father, the Son and the Holy Spirit. The worst triad is; Satan, sin and self.

This positioning of the self as a 'bad guy' is hard for most of us to accept. Many will ask, 'Well then, who am I?' Actually, you are somebody of extreme importance, so much so that it is difficult to grasp. How can a person grasp that he or she, as a human being, is the most important of all of God's sublime creation?

In order to discard the demands of the self and self-importance and to put one's God given identity in proper perspective it is necessary to study the Bible especially the New Testament. In the final analysis, there is no substitute for this in order to get a firm foundation and understanding of the 'wonder of being human'. That is, the study of the Bible is a categorical imperative.

Only by this means, usually aided by someone well founded in the scriptures can one really distinguish between *self* and identity.

Each individual is of utmost worth to Almighty God and is destined for eternal bliss; but only if one is devoted to God and His Christ (Jesus).

This concept is difficult because of the 'pride of man' that is, of every solitary one of us humans! Until the realization of how filled with false-pride each one of us is – the progress will be slow. It is very difficult to grasp just how unique and invaluable to God we are, without the old self, but with our new individual identities.

The Bible infers that we will still know each other and be able to be together. ** It is important to understand, we are each one endowed with an eternal and unique identity, so that we are somebody – every one of us – <u>and yet all are one in God</u>

It is hard to know this about who we will be and yet not long for our old carnal self, much the same, as it is not easy to give up old bad habits such as smoking. No matter how grand this is to contemplate, we *will be free of pride*!

God is the one who could feel pride about His creation; however, the proper term for God is **glorified by** His creation. **We** (you and me) **humans** are the only part of His creation, the only one of His creatures that can **glorify Him**!

We are living in a materialistic, Godless, lost, society wherein if it paid to be good, then it would only take intelligence to be good. **Intelligence is not the essence of goodness.** Goodness is spiritual and is of faith, and faith is demanded by God, Goodness leads to love which unites one with God, Who is eternal. To be one with God is joy and peace forever. It is the eternal destiny of the believer.

Ever since our birth there has been a struggle between God's creation of the soul (the host), and the devil's false construct of the ego or sinful self (the parasite). This parasitic process has been taking place to varying degrees in different individuals since birth and leads to death and damnation if not brought under control. This was brought about by the 'fall' of man in the Garden of Eden.

Once we accept 1) Jesus Christ as the Son of God, and 2) that we need Him to rescue us from Satan, sin and death. We are given the Holy Spirit to comfort us, strengthen us and sanctify us. The Holy Spirit enables us to overcome, to a large degree our old self or ego and begins a process of rebirth, wherein **our God given soul** is aided in becoming the center of our identity.

In other words, we can then 'die' to the sinful self (ego) and be alive in Jesus Christ and thus acceptable to God. Thereafter we are less and less contaminated by the devil and sin. Our soul is revived, strengthened and becomes increasingly dominant over the ego (i.e. the sinful self). We are then eligible for eternal life with Jesus Christ and God, in Heaven.

The more illuminated by God the more shadowy the ego becomes, until Jesus comes to get us. Then the full light of God pervades us and we are completely free of even the shadow of the sinful self. Satan, sin and death, are completely exterminated from our souls. Our mind and soul are spiritualized, our bodies are glorified we are filled with the Spirit of God and are immortalized. We then live in the eternity of Heaven with Almighty God the Father, the Son and the Holy Spirit.

It is important not to judge anyone else. Even as the Pharaoh was used of God to accomplish His purposes with the Israelites, in what appears to be a similar way He may be using someone you judge unworthy of Heaven! There are indications in the Bible that God wants most all men to be saved; thus, we do not even know with absolute certainty how He will deal with such humans. (Authority for this, 1Timothy 2:3-6)

*   (from, 'Elegy Written in a Country Churchyard', by Thomas Gray)

** Matthew 17:3 for example

'My God! my God! look not so fierce on me!
Adders and serpents, let me breathe a while!
Ugly hell, gape not! come not, Lucifer!
I'll burn my books!-Ah, Mephistophilis!*

XVIII
THE TRIBULATION

A concept that is held by many Theologians is that we (the chosen) will be called out from the unsaved either before or during the Great Tribulation. The Great Tribulation is that period of seven years that will mark the end of life, as we know it as well as the earth as we know it.

For example, the passage in The Revelation (of St. John) Chapter 4 verse 1 where the voice said. "Come up here," is taken to mean the saved will be taken into the Heavens and the unsaved will be left to undergo the tortures of the Tribulation of seven years on the earth.** This viewpoint is set forth in the series of best selling books called, 'Left Behind.'

According to these books, the saved (chosen or elect) will be spared the horrors of the Tribulation. The belief or inference is that the balance are damned if they undergo the Tribulation. This could leave many feeling they have been damned if they are not among that first wave that are 'taken up'

However, the Bible states that countless saved ones (chosen) will undergo all or part of the Tribulation. The *viewpoint* espoused by Christ Himself (in two Gospels and in Revelation) indicates many (too vast to count) will go through the entire Tribulation.

For example, Christ Himself informs us in Matthew 24: 4-51, especially verse 31 which states; "And He will send forth His angels with the sound of a mighty trumpet blast, and they will gather together His chosen ones from the farthest ends of the earth." Mark Chapter 13:5-33 especially verse 27 "And He will send His angels to gather together His chosen ones from all over the world- from the farthest ends of the earth and heaven."

The context of both of these references indicate that the 'gatherings' will be at the end of the Great Tribulation. A further study of Revelation yields the following references to the gathering of God's chosen at the beginning of and during the Tribulation, for example: Revelation 6:9-11 (martyrs); 7:3-4 (Jews); and 7:9, & 7:13-14 (countless numbers); all show these coming out of The Great Tribulation [even its worst part]) .

These Bible excerpts will be of much relief for many who will still be around during the Tribulation, as they plainly reveal that the gathering of His chosen ones will be during, and at the <u>end of the Tribulation</u>! ***

This is a most important point because if we are living in the End Times and we begin to see the Tribulation occurring all around us, we may well believe, falsely, that we have not been saved because we have not been taken up!

This drives home the point that it is very important that we each one of us study the Bible on our own and pray, because the Holy Spirit will speak to each one of us that are seeking and interpret the Bible for us. This is not to say that it is not helpful to seek teachers of the Bible or to have study groups with other laypersons.

In all of this, it is vitally important to remember, **hope is necessary**. That is, if you do not feel you have faith, but you want it and wish you had it. It seems apparent from the Bible that *HOPE IS NECESSARY* and you will be given faith. Remember hope is one of the three prerequisites God requires of us, i.e., faith, hope, and love. (1 Corinthians 13:13). Love is required, hope is essential and faith is a gift, which you may not have received yet!

Those who hate God and do not want anything to do with Him are the ones who are lost (damned, unsaved) in the final analysis. These people use God for their own purposes, deliberately and prefer to think of a world without God. They deny any help offered to them to see things differently.

They try to destroy anything that is of God as well as anyone that promotes our loving God. Love to them is nonsense and living the moment is what is of ultimate value – even at the expense of others. Anything bad that happens they lay the blame on God. They deny that Jesus was the Son of God and attribute the works of the Holy Spirit to the devil. (See Mark 3:29).

* (from, 'The End of Faustus', by Christopher Marlowe)

** come up here was addressed to John only, it has no references to other Christians.

*** authority, Matthew 24:30-31 & Revelation 7:9-14, and 20:4)

The Moving Finger writes; and, having writ,
Moves on: nor all your Piety nor Wit
Shall lure it back to cancel half a Line,
Nor all your Tears wash out a Word of it.*

XIX
JUDGMENT

Where is Heaven? The Bible indicates that it is right here on earth except it will not be the earth, as we know it. It will be a new heaven and a new earth. Authority for this 2 Peter 3:13 and Revelation 21:1-8.

This will happen after the Millennium and after the final judgment. A reading of Revelation Chapter 20 would appear to indicate that some saved ones will not be brought back to life until the final judgment (verse 12-13), in other words they will not reign with Christ during the Millennium.(verses 4-6).

Chapter 20 could indicate by verses 14 & 15 and Chapter 21 verses 6 & 7 that many people will be saved from eternal damnation. However, Chapter 21 verse 8 is a final warning to those who are teetering on the edge.

As one studies the book of Revelation, especially Chapters 20 & 21 in their <u>entirety,</u> it appears that there are important distinctions to be made:

1) those who are going to live eternally and blissfully in the New Heavens or the New Earth;

2) and those who will be assigned to <u>eternal damnation</u>.

We just do not know who will go into the Kingdom of God and who will not. For example, In Matthew 21:23-32, Jesus told the Chief Priests and the Elders (of the church) that the publicans and harlots would go into the Kingdom of God before them. Please also notice that Jesus did not say that the Chief Priests and Elders would not go into the Kingdom of God, but rather that harlots and publicans (Jewish tax collectors for the Romans, many of whom were dishonest and many were hated.) would go into the Kingdom of God first.

Another point to make is that we do not know, of the ones who are not destined for the Kingdom of God ** all go to eternal damnation. Only God knows this. Furthermore, we have no legitimate grounds upon which to stand and criticize (or reject) God. We just do not know enough and we do that at our peril. In fact, it is senseless and ignorant to do this and thus condemn oneself.

God is a God of love, why would He unjustly condemn anyone. In the final analysis, to say God is unfair, unreasonable or heartless is just an excuse to reject God, which one can do! But why? The only thing to be gained by taking this ignorant position could be the loss of paradise, and eternal separation from your loved ones!

Was Judas Iscariot sentenced to eternal damnation? If we do not know the answer to this would it not be time to give up on judging others? This includes excessive worrying about loved ones who appear to reject everything that is said to them about: a) receiving Christ or b) going to Church or c) reading the Bible.

Remember, we can pray for them and try to be a good example of one who loves God. In which case the beloved one may very well notice things about us that raise his/her admiration and wonder, which could create enough interest to cause further searching. God has ways of working without our interference!

It is easy at times for one to feel that God is unfair, unjust or even allows evil to happen needlessly! This is Satan at work wanting a person to reject God! There is no doubt that it stretches our credibility at times. It is so very important to really struggle against this effort by the devil. Whether we believe it or not there is a very real Satan and evil that does exist outside of God.

Denying Satan's existence is part of the devil's plan. This is why in Hebrews chapter 11 verse 6, God declared, "...it is impossible to please God without faith. Anyone who wants to come to Him must believe that there is a God and that He rewards those who sincerely seek Him."

What needs to be accepted even if it is not understood is that there is a reason for this powerful counterforce to God, which is called Satan or the devil. There are cosmic and psychic reasons for the existence of evil, but the devil (Satan) and evil will not exist forever. The authority for this is Revelation 20:10

God must hold back and let evil exist to a certain degree for a predestinated period. This is necessary (for an undisclosed reason) so that we will be able to exist with Him in eternity. The problem involves the almost total freedom God has given each one of us; we are usually unaware of the awesome responsibility, which that places upon all of us.

In this God-given freedom, it is inevitable that humans will make mistakes and create problems for which some 'innocents' pay the price. This fact and the devil's work, make the earth for some, a tragic place, and a place of suffering or sorrow.

An attempt at an inadequate and partial analogy that gives an inkling or hint of what is involved follows. That is, why God does not destroy the devil outright.

Let us assume for the sake of this analogy that an electronic engineer represents God and viruses represent the devil. In addition, assume that computers represent ordinary people.

Viruses (the devil) attack and make inoperative computers (people). There are two solutions,

1) The electronic engineer (God) could destroy the computers (people) or

2) Destroy the viruses (devil).

However, the electronic engineer (God) refuses to destroy the viruses (evil) immediately but rather changes the computers (people) and makes them resistant to all viruses (evil) not only present and past but also future.

Thereafter computers (people) can live in perfect harmony with the electronic engineer (God) free of any future possible invasion by viruses (Satan and his devils, which will no longer exist).

Thus, in God's time the world as it is, works for the benefit and final reward of all. In other words, the result will be to our eternal good and God's glory! For this to come about for any human individual requires faith, an enduring faith.

* (from, 'The Rubaiyat of Omar Khayyam' by Edward Fitzgerald)

** there is hope even for those who are headed for eternal damnation. For example, Christ in one of His last acts before His crucifixion forgave a criminal on the cross next to Him. This surely shows the boundless mercy which God offers. Nevertheless, it would be playing an extreme game of 'Russian roulette to hazard this last breath conversion!

For now we see through a glass, darkly;
But then face to face:
Now I know in part;
But then shall I know even as also I am known.
And now abideth faith, hope, charity,
These three;
But the greatest of these is charity.*

XX
THE CHURCH
FRACTIONATED

Darwinism and Freudianism, although they stimulated further study, both tried to claim too much. Some of their followers tried to suppress any further real advances that might disagree with some of their unproven theories. This is also true for medicine, astronomy and all human endeavors and advances, including religion.

Unfortunately, once a theory is developed that has some truth to it or can explain some problems; the originators and/or their followers often attempt to resist any new or different viewpoints. They try to make their hypotheses and theories very hard to dislodge. (Possibly, they do this for reasons of power, money, or fame). This creates great difficulties and roadblocks to further advances of science or being open to further truth

For example, as stated previously, Ptolemy believed the earth was the center of the world and that the Sun revolved around the earth. Copernicus came along and said that the earth revolved around the sun, this was accepted by Catholicism as long as it was held to be a <u>theory</u> only.

However, when Galileo came along and said that it was a <u>fact</u> that the earth revolved around the sun, he was taken to task by the Church, made to recant, and placed under house arrest.

The reason for this resistance to and even persecution of the truth is the fallibility of man. Man is subject to pride and often tries to maintain his exalted position stemming from an advance in knowledge he has achieved.

Another reason for this opposition to change is that some have stakes in lying and suppressing the truth. Different reasons motivate men to try to maintain the status quo such as racism, accumulating wealth, and fame, political or ecclesiastical power.

The result is often humanity going off on a tangent and not only getting further away from the truth but being resistant to it. On the other hand, the practitioners of the Fifth Dimension advanced truth and were not adverse or opposed to further revelations of the truth.

The prophets such as Abraham, Moses, David, and John were all forerunners of the ultimate truth in Christ Jesus followed by Paul who explained the truth of Christ to the Gentiles.

That is not to say that some theologians are not guilty of the same thing. As one example of this resisting the truth, is the strongly held concept of the 'pre-tribulation rapture' position, which some theologians and writers hold. This was previously described, in the position set forth by the writers of the 'Left Behind' series. Some fundamental pastors also cling to the 'pre-tribulation rapture' view.

This despite the fact that Christ Himself stated plainly in Matthew Chapter 24, Mark Chapter 13, and Revelation Chapter 20 that His chosen ones would be called to Him at the end of the tribulation, both the living and the dead. This directly opposes the 'pre-Tribulation Rapture' theory, which states the chosen ones will all be taken up to be with Christ (raptured) before the Tribulation begins!

Although this is neither the major, nor the worse inaccurate stance taken by some theologians and their flocks, it is a current example. These theological differences have always led to, and still do, arguing, judging, expelling and even killing people that hold a different interpretation of the Scriptures! Some views have been much more pernicious as in the 'Roman Catholic Inquisition'

There has always existed an un-Christian attitude, on the part of a small and extreme minority, which can bring about rejection of the Bible, Christ and the Church by non-believers. It is time for <u>all</u> so-called Christians to become more loving!

Another problem the Christian churches have had is arguing, which can lead to the breaking up of Protestant Congregations into many different sects and denominations and even excommunicating dissidents.

Much more important however, is the widely held Darwinian, Godless, position that our universe is all a chance happening. This is affecting our schools and pervading our children's lives and consequently the mores and morals of our society to an alarmingly increasing degree.

* (1st Corinthians:13, 'The Bible Designed to be read as living literature', by Ernest Sutherland Bates.)

"In the beginning was the Word,
And the word was with God,
And the Word was God.
The same was in the beginning with God.
All things were made by Him;
And without Him was not anything made
that was made.
In Him was life;
And the life was the light of men.
And the light shineth in darkness;
And the darkness comprehended it not." *

XXI
THE MAN OF THE FIFTH DIMENSION: GOD-MAN

There was, is, and will be one God-Man. He was Christ-Jesus born of a woman sired by God. Who was He? He was the fleshly reflection of God and one of the Trinity.

What is the Trinity? The Trinity is Almighty God Who is one, having three reflections: Creator, Man, and Spirit. These three are usually referred to as Father, Son, and Holy Ghost. The

Son was pre-existent with God and once dwelt among us (humans) as one of us.

Humans, male and female, were created to be reflections of Jesus Christ. Tragically, the first god-man (Adam and Eve) were corrupted by Satan. The original God-Man, Christ, the prototype of man and woman, although taking human form in Jesus of Nazzareth was, remained and always will be perfect.

We inherited contaminated genes from Adam and Eve; this contamination was a susceptibility to the temptations of the devil. Christ was pre-existent to Mary and was merely 'channeled through her, with God (Father, Son, and Spirit) all things are possible.

There are great mysteries, for example, man-woman (remember as God said) become one in marriage, He also said that we (man and woman) become the 'bride' of Christ! The greatest mystery is the origination and existence of Satan (the devil) and evil.

We are creatures, and even though the most exalted ones in God's creation, He remains our originator and <u>the</u> creator, not we of ourselves! He has promised that someday we will know, even as we are known. (authority, 1Corinthians 13:12)

We are to praise, worship, and glorify Almighty God of Whom the most tangible, visible, and knowable representation is Christ Jesus.

We can be so sorry for the sins we have committed against God and Man, but our 'sorriness' can never undo or change this or how much we have grieved the Spirit. **But,** hallelujah, Christ could and did (and will) take all those sins upon Himself and thus make us, not only acceptable, but also perfect in God's sight. Thank you Jesus!

Furthermore, Christ has gone on to completely defeat the devil and his minions, and this will be forever (authority Revelation 20:10). He (Christ) now resides eternally at the right hand of God, and He continues to intercede for us until we enter eternity (where we will be perfect).

As for the devil and his evil angels, they will be allowed to become active again, briefly, after the Millennium. Much as a snake will writhe around until sunset before it dies (or so the old wives tail goes). This is analogous or indicative in our limited understanding.

Jesus Christ is the hero of all heroes. He is greater and more to be admired and **loved** than any human that ever existed or than any mythical god or creation of man's imagination.

His sacrifice was so far beyond any possible human comprehension. Just imagine the indignities, pain, and injuries He suffered at the hands of <u>His creatures</u>! Even so, He loved us, without measure, despite how besmirched and selfish we had become.

Despite our experience and/or observation of how evil our world is, God (in all of His forms or reflections) is a **God of love.**

Jesus Christ the perfect Man, the Man of Love who was without fear or hate and yet expressed all of the healthy human emotions. He was and is the perfect Man and true God.

He has forgiven every sin known to man except one and that is the sin against the Holy Spirit, but what is that? Anyone who has read this far, has not, committed the unforgivable sin. The unforgivable sin is, denying the Holy Spirit, consistently, and permanently after He is within that person.

The Holy Spirit is within a person who has accepted three conditions:

1) Jesus Christ was the holy begotten Son of God, Who died for the salvation of sinners, and was resurrected to eternal life in Heaven with God,

2) He or she is a sinner inextricably mired in sin and without hope, outside of Christ,

3) He or she repents and continues to try to change his/her life with the help of the Holy Spirit.**

It is important to remember that we all backslide and fail from time to time and even give up, we may even reject God or Christ. The vital thing is that we reconsider and re-establish our faith. This is confirmed by the fact that our faith again begins to grow, little by little.

This is irrefutable proof that faith is a gift which we cannot earn and do not deserve; but which we have because God is a God of love and his patience and forgiveness is limitless in Christ.

* (from, John 1, 'The Bible Designed to be Read as Living Literature', by Ernest Sutherland Bates)

** (Authority for these statements is 1Timothy 2:4, Matthew 12:31 & 18:21 - 35.

But with unhurrying chase,
And unperturbed pace,
Deliberate speed, majestic instancy,
They beat-and a Voice beat
More instant than the Feet-
"All things betray thee, who betrayest Me." *

XXII
A SINNER – A SAINT

At eight years as an active, robust boy, I was struck down with an inflammation of the spine, a 'boil' developed right over a kidney. Doctors said they had to operate but were afraid of puncturing the kidney, so they held off as long as they dared, hoping it would abscess and drain.

It turns out that the diagnosis was tuberculosis of the spine for which there was no known cure in 1930. There was not much hope for my life. The time came when the doctors could wait no longer; they scheduled the operation for the next day.

That night my Aunt Adina came to our house and lay on my bed for over two hours praying (she was a Christian Scientist). She left around ten or eleven P.M. At midnight, I woke up and found that I was lying in a pool of pus! The boil had come to a head and burst! I got up and went to the head of the stairs.

I yelled down to my parents, "God did it! God did it!" The doctors did not have to operate. It took two years out of grade school for me to recuperate. One of which I made up through tutoring.

During my convalescence, religious aunts and grandmothers were my main visitors. I became a born-again Christian, however, when I reentered school I was a 'goody two shoes' and was 'picked on' by the other fourth graders. This, not really 'fitting-in', went on until my second year in college. One afternoon while reading the Bible, I became 'fed-up' with it all. I threw the Bible across the room. I had had enough, so

I changed my life style. I began smoking and drinking alcohol.

After I entered the Naval Air corps in 1942, I began dating promiscuous women. By the time of my discharge from the Navy, I had begun regretting my rejection of the Bible. I began to search for my discarded faith.

Altogether, it took seven years before God granted me the privilege and wonderful gift of faith again! This came at an evening service of a small country Baptist Church where they were showing a Billy Graham film. Once again, I began attending church and was very thankful to God for giving me faith again!

However, after many years, this was not to last. For the first time I began an illicit but complete and total relationship with a woman, whom I will always love. Thereafter I got a divorce. After some time of so badly grieving the Holy Spirit, I began repenting of what I had been doing and altered my behavior. Our wonderful Lord took me back again.

Despite all this and more God loved my soul in Christ, Who not only redeemed me from the devil but <u>also redeemed me **from my self!**</u> That was years ago and although by no means without sin, (not being perfected yet), my life is one of a close and satisfying relationship with the Holy Spirit.

I had an ecstatic experience once when I was in my thirties. I was on my knees praying and suddenly I was in a state of complete euphoria. I did not know what was happening!

Unexpectedly, for no reason, I heard myself saying, 'Take my hands Lord', and held my arms up. Years went by and that experience made no sense until now. For the past seven years, I have been confined to a wheel chair, unable to walk.

Over the last several years, this is the fourth book I have written, all on and about faith! Now I know why I said, fifty years ago, 'Take my hands Lord.'

God forgave David, and although I am no David, yet every one of His children is so precious to Him, even me, that He forgives us repeatedly. What a wonderful Savior that would die to save people like me!

Therefore, sinners are not beyond the reach of Christ; however, one needs to reach out to Him. Christ never forces anyone but no one can sincerely reach out to Him until pride is gone or at least going.

Honestly reaching out to Him can be seriously impeded if:

1) One has never done anything 'really bad', (See page 60 and 61)
2) One is rich, powerful or generally successful.

A life and a poem that gives beautiful testimony to the truth of God's love and forgiveness is; 'The Hound of Heaven' by Francis Thompson. It is well worth the effort to go to the library and read this.

He was a drug addict that was finally taken in by a convent in the last years of his life and nursed back to health. His was a wonderful talent of putting words together in a pleasingly harmonic way that had a powerful message. God never rejected him and finally he came to Christ. Sin, what is it? It is not trusting the Lord with your life, all of your life.

"So we have stopped evaluating others by what the world thinks about them. Once I mistakenly thought of Christ that way, as though He were merely a human being. How differently I think about Him now! What this means is that those who become Christians become new persons. They are not the same anymore, for the old life is gone. A new life has begun!"   (2 Corinthians 5:16-17)

*   (from, 'The Hound of Heaven' by Francis Thompson)

"O Captain! my Captain! our fearful trip is done,
The ship has weathered every rack, the prize we
sought is won,
The port is near, the bells I hear, the people all
exulting,
While follow the eyes the steady keel, the vessel
grim and daring;
But O heart! heart! heart!
O the bleeding drops of red.
Where on the deck my Captain lies,
Fallen cold and dead." *

XXIII
CRUCIFIXION AT CALVARY

On a dark day with lightening flashing, and
dark storm clouds shedding rain, suddenly the
veil in the inner sanctum of the Temple was
split from top to bottom. God was hanging on
a cruel cross on Calvary.

He had been mercilessly lashed, beaten, spit
upon, mocked and crushed under the weight of
carrying His own cross up the hill to Calvary.

The Jews and the Romans hated Him and made merciless sport with Him.

Before His agonizing death He called out to His Father, "Father, forgive them for they know not what they are doing!" He is still calling out today for men and women as we crucify Him repeatedly.

Selfish materialism, promiscuity, pornography and perversions, violence, hate, and murder, are worse rejections of Him than His killer-tormenters displayed. Because today it is known, who He was/is and there is the advantage of hindsight. Sadly, foresight appears to be lacking, in the multitudes.

So many of us are blind and do not know that we cannot see, just as we are deaf and do not know we cannot hear.

"Come now, let us argue this out," says the Lord. "No matter how deep the stain of your sins, I can remove it. I can make you clean as freshly fallen snow. Even if you are stained as red as crimson, I can make you as white as wool.

If you will only obey me and let me help you,..."
Isaiah 1:18-19...

He forgave me, He will forgive you of all your sins

I began this book in 'fear and trembling'; I finish it with relief and freedom! Free at last in my Lord, Savior and Redeemer, Jesus Christ!

My wish and purpose for writing this book was to give hope to others.

The author, Bruce P. Burns

After you accept Christ, the Holy Spirit begins to teach and to change your spirit, which then becomes dominant over the 'ego' and the 'self'. Your soul is your real identity, who you are, a unique individual, created by God.

* (from, 'O Captain! My Captain!', by Walt Whitman)

XXIV
PROFESSION

If you are not certain of your salvation, it is a simple procedure to receive Christ as your savior, at the same time securing your permanent acceptance by God.

Read the following, sign it, preferably in the presence of someone else

I want to be with my loved ones in eternity living a life wonderful beyond that I can imagine.

I need the forgiveness of Jesus Christ and I want it. I believe that He is the Son of God and that He died on a Cross at Calvary for me and all other sinners.

Take my life Lord and help me to live a life pleasing to you and obedient to the New Testament.

Signed:_____

Dated:_____

Witness: _____

The Lord bless thee and keep thee:
The Lord make his face shine upon thee,
And be gracious unto thee:
The Lord lift up his countenance upon thee,
And give thee peace.

(The Bible, Numbers Chapter 6)

www.ingramcontent.com/pod-product-compliance
Lightning Source LLC
Chambersburg PA
CBHW022018170526
45157CB00003B/1274